Exploring Perspective Hand Drawing

Fundamentals for Interior Design 2nd Edition

Stephanie M. Sipp
with Cheryl L. Taylor

SDC
Publications

SDC Publications
P.O. Box 1334
Mission, KS 66222
913-262-2664
www.SDCpublications.com
Publisher: Stephen Schroff

ISBN-13: 978-1-58503-901-2
ISBN-10: 1-58503-901-2

Printed and bound in the United States of America.

Contents

TEXTURE & PATTERN 79

SHADE & SHADOW 97

FURNITURE 113

INTERIOR DETAILS 151

REFERENCES   **173**

Preface

This companion *Activity Sketch Book* provides a step-by-step approach for practicing and developing basic line drawing techniques. I recommend having your *Perspective Hand Drawing* reference book available as you work through the activities; they correspond directly with the information in the book, which provides additional support material. In addition, a number of the enclosed activities have a corresponding video to demonstrate the techniques described in the book. A CD icon at the end of the instructions denotes these activities. Two video clips that demonstrate adding furniture blocks to a one-point grid and a two- point grid support the design projects.

The collection of activities in this book are intended to provide a foundation for developing basic line drawing techniques. These skills are brought together in Chapters 8 (One-point Perspective Project) and 9 (Two-point Perspective Project) of the reference book, where you will find instruction for creating complete interior design projects. For these projects, you will be using drafting supplies and large sized paper; therefore, they are not included in the activity book.

Some activities have examples of objects that were drawn by me and yet you are asked to find your own object to use as a subject. Copying my image will not provide you with the same practice as drawing from an object in front of you. In addition, I encourage you to adjust or change the activity to meet your own drawing needs.

Below is a quick list of the typical supplies and tools you will need to complete the activities. Please refer back to Chapter 1 (Orientation) of the *Perspective Hand Drawing* book for more detailed information on using the tools.

BASIC DRAWING SUPPLIES

Drawing Pencils	HB, B & 2B wood drawing pencils one each
Eraser	White block eraser
Sharpener	Hand held, small pencil sharpener
Drawing Markers	Thin, medium, wide drawing markers in black or sepia
Eraser Shield	Small metal piece with holes used in erasing
"T" Square or ruler	12" small, plastic "T" square or ruler
Triangle	90 degree small drafting triangle

Below is a list of drafting supplies need to complete the one-point and two-point perspective projects outlined in Chapters 8 and 9 of the *Perspective Hand Drawing* reference book.

BASIC DRAFTING TOOLS

Roll of Tracing Paper	Least expensive, white or yellow 24" wide tracing paper
Drafting Tape	Low adhesive drafting tape or dots
Architectural Scale	Triangular shaped or flat architectural scale
Triangle	An 8 inch or larger architectural triangle
T-square 36	Metal or wood 36 inch T square
Drafting Surface	Drafting surface large enough for a 24"X 36"paper
Velum Paper	Plain 24"X 30"velum paper

Chapter 1

Orientation

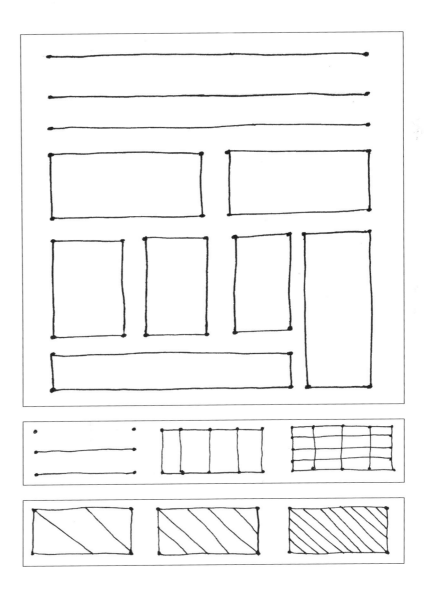

ACTIVITY 1.1

One of the best techniques for starting to practice the skill of drawing is to use guide dots or guide lines. By anticipating where you want your line to go and putting a dot at that spot, you can use this as a target for your pencil to move forward.

In this activity, practice drawing consistent lines using the guide dots below. Lines that are the same weight and thickness from start to finish.

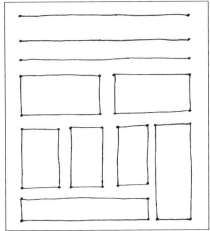

❶ Draw lines and box shapes that are similar to the drawing on the right using the dots below.

❷ Put your pencil on the left dot and draw a line to the right dot.

❸ Let your eye move forward to the dot on the right. Your hand will know where to go as you look ahead to the destination dot.

❹ Repeat these steps to draw two more lines.

❺ Use the same dot-to-dot technique as you draw the rectangular shapes to finish the activity.

ACTIVITY 1.2   LINES WITH WEIGHT VARIATION

Practice the same technique of starting with your pencil on a dot and moving it toward the next one. This time, practice drawing simple squares and rectangles and dividing them with lighter value lines. Vary your line weight by using a darker line for the contour of the box and a lighter line for the pattern.

❶ Using a light pencil line, draw six additional rectangular shapes the same size as the shapes below. Plan to use your T-square to assist with drawing these shapes.

❷ Darken the outside of the box shape and erase any extra guidelines.

❸ In each row, using a lighter line, reiterate the line pattern provided on the left.

❹ When adding the diagonal lines, start in the top left corner. Practice using your eye to draw lines that are equal distant apart.

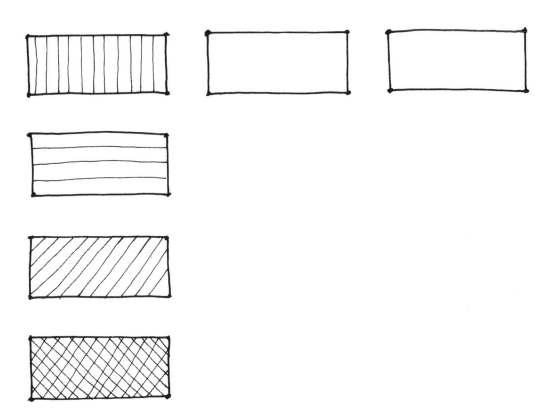

ACTIVITY 1.3   DRAWING FURNITURE IN PLAN VIEW PART I

In this activity, you will start with practicing drawing furniture shapes. Each of these shapes use guide points and guidelines to assist with drawing the shapes. After this, you will be drawing these furniture shapes in plan view. Refer to chapter 1 of the reference book for an illustrated example of this activity.

❶ In the first row, use the dots to draw horizontal and vertical lines creating three separate square shapes.

❷ In the second row, draw the square on the second box draw an "X" from corner to corner with a dashed line. Use this "X" as a guide to draw the inside of a chair arm. Repeat this in the third square.

❸ In the third row, use the first rectangular shape and dashed lines to draw inside line of the sofa arms and back. Use the center of the "X" to draw the pillow line. Repeat these steps in the second rectangle.

❹ In the fourth row, use the square to draw a circle. Start with the "X" and divide the square. On the "X" divide each line into 1/3 with a dot. Use the top third dot as the guide for your circle along with the point where the cross intersects with the square.

Note the next page has step #5 for this activity.

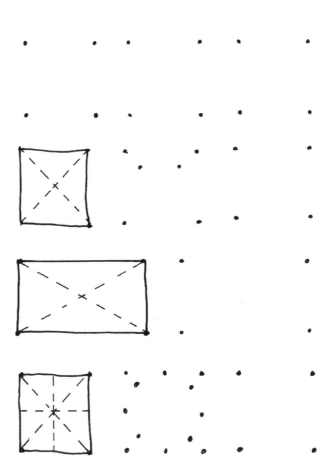

ACTIVITY 1.3   DRAWING FURNITURE IN PLAN VIEW PART II

❺ In the rectangle below, draw a plan view of a seating grouping. Plan to use the steps that you just completed. Refer to chapter 1 of the reference book for an illustrated example of this activity.

ACTIVITY 1.4   VALUE GRADATION WITH PATTERN

Using a design drawing style, draw a gradation of value with variety of pattern. Refer to chapter 1 of the reference book for an illustrated example of this activity. Plan to imitate the style of the patterns found in the textbook.

❶ For each row, select a pattern that can be used to draw in a gradation of value. Fill each square by having your pattern touch the sides.

❷ For each row, start in the first block with the lightest value and build in each block to the darkest value. You will have more space between your pattern lines at the beginning with the pattern lines coming together as you draw each block.

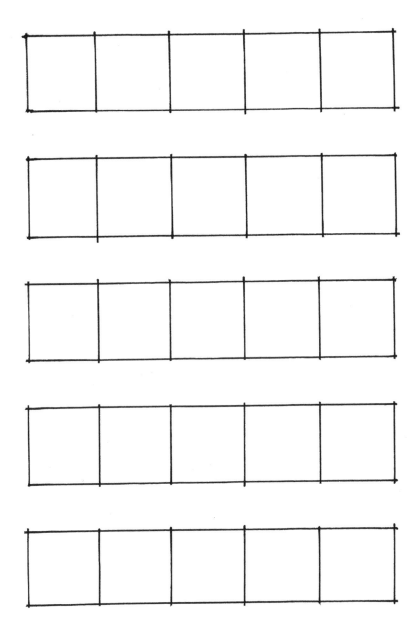

In this activity, use the rectangular shapes to show contrasting values in a design drawing style. Refer to chapter 1 of the reference book for an illustrated example of this activity. Plan to imitate the style of the patterns found in the textbook.

❶ In the first row, draw lines and dots, to create high contrast patterns in each box with a low value pattern against a high value pattern.

❷ In the second row, draw patterns, to create a medium contrast in each box.

❸ In the last row, create your own value combinations.

In the first part of this activity, you will be drawing different shapes each starting with a rectangles. Add value in gradation to each shape. Refer to chapter 1 of the reference book for an illustrated example of this activity.

❶ In the first row, use the four dots to draw a rectangle shape. Draw a gradation of value from light to dark with dots and circles in each shape.

❷ In the second row, use the formula shown to draw three oval shapes. Draw a gradation of value from light to dark with broken lines in each shape.

❸ In the third row, use the formula shown to draw three triangles. Draw a gradation of value from light to dark with lines in each shape.

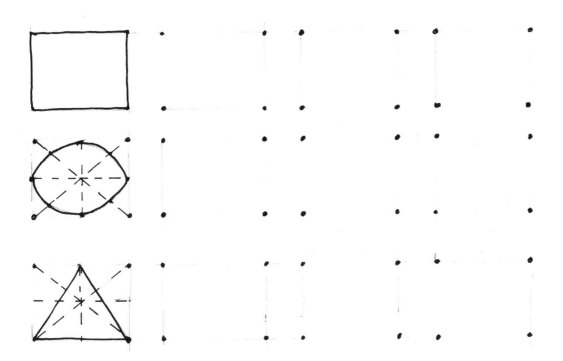

In the second part of this activity, you will be using shapes to create four compositions from the shapes that you have been practicing. Plan to use the elements and principles of design when you are putting this together. Please plan to create your own composition. Refer to chapter 1 of the reference book for an illustrated example of this activity.

Chapter 2

The Box

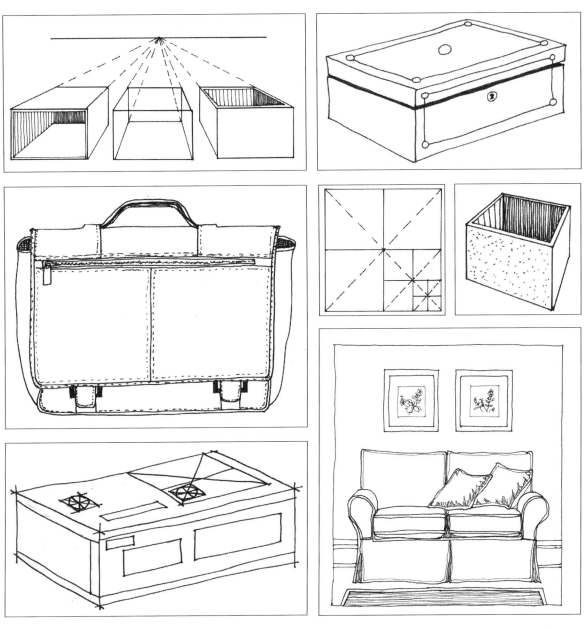

In this activity, use the drawing below to complete the one-point perspective box shape. Notice the starting point is the flat front of each box. You will add the rest of the box shape using the vanishing point. Your final drawing should match the example.

❶ In pencil, for each box, use your straight edge to extend the dashed perspective lines to connect the vanishing point to the flat front of the box. This will create the top and sides.

❷ Draw the back of the box with your straight edge, creating lines that will be parallel to the flat front. Darken the box lines.

❸ Define the inside of the box using two lines for the edge and a vertical line for the inside corner line. Add additional vertical lines on the inside of the boxes as shown in the example.

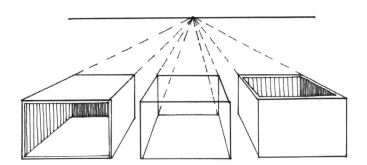

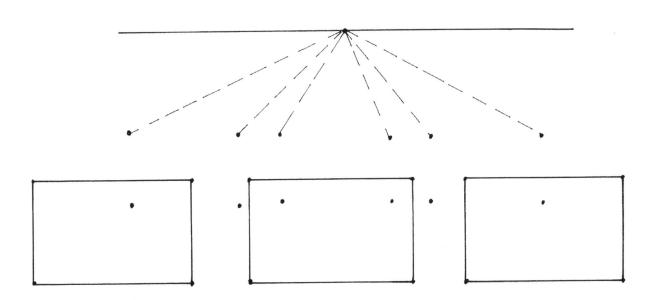

In this activity, follow the steps below to add one-point perspective box shapes similar to the example.

❶ Start your first box shape above the horizon line provided below. Draw the box shape that represents the flat front of the box.

❷ With a straight edge, draw dash lines from the vanishing point to each corner of the box.

❸ Determine the depth of the box and then add parallel lines to define the back of the box.

❹ Redraw the lines of the box to darken and to make a distinction between solid lines and dashed lines.

❺ Repeat these steps, adding another box below the horizon line and another on the horizon line.

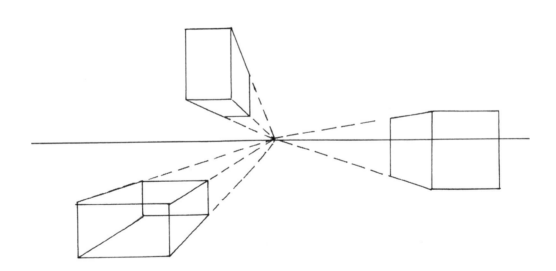

This activity is similar to activity 2.2, however, this time draw your own horizon line with boxes above and below as shown in the example. Use the directions on this page and complete your activity on the next page.

❶ Before starting your drawing, add guidelines to the drawing below using your straight edge. These guidelines should extend from the vanishing point to every corner of the box. Notice how the angle for the sides of a one-point perspective box always comes from the single vanishing point.

❷ On the next page, start your drawing with a horizontal line across the middle of the page and then mark the vanishing point in the center of the line.

❸ Draw a flat front box above the horizon line.

❹ With your straight edge, draw dashed perspective guidelines from the vanishing point to the corners of the flat front box.

❺ Determine the depth of your box and add lines that are parallel to the top and side edges of the flat front box to define the far end of the box. Darken the contour of the completed box shape.

❻ Continue to add more boxes from different positions above, below, and on the horizon line. Remember to start each box with the flat front.

❼ Draw the inside view of several box shapes.

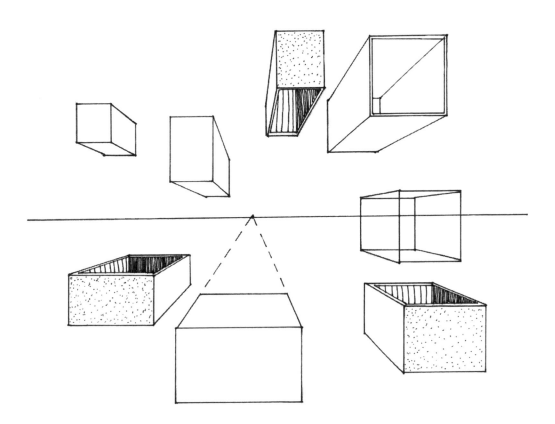

ACTIVITY 2.4   OBJECTS IN ONE-POINT PERSPECTIVE

For this activity, start by finding a simple rectangular shaped object to look at as you do your drawing. Face the object so there is a flat front. Look carefully at the proportions and the details of your object. Taking time to notice the details of your subject is an important step for creating successful drawings.

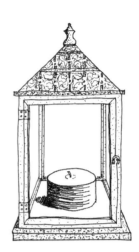

❶ In the space below, start the drawing with a rectangular box shape at the bottom of the page. Plan to make the flat front shape at least 2 or 3 inches wide. Use your straight edge tool for the first four steps.

❷ Mark a single vanishing point above and in the middle of your box shape.

❸ From the corner of your flat front shape, draw the top perspective lines angled toward that single vanishing point. These lines form the top of your object.

❹ Use your eye to determine the depth of your object and draw a line parallel to the top line of your box. This defines to back edge.

❺ Plan to draw the details of your object free hand without the straight edge tool.

ACTIVITY 2.5 TWO-POINT PERSPECTIVE BOX

These activities focus on two-point perspective with four box drawings to complete. Use the drawings below and on the next page to draw two-point perspective boxes. Notice in the example with the numbers that the two vanishing points are off the page.

❶ Start with a line connecting dots 1 & 2 to form the leading edge of the box.

❷ From dot 3 to 4 draw a line parallel to the leading edge. Do the same from dots 5 to 6. These lines define the back edges of the box.

❸ Darken the perspective lines between 1 & 3, 2 & 4, 5 & 1, 6 & 2 to complete the right and left sides of your box. Notice that the top and bottom perspective lines are coming from the left and right vanishing points that are off the page.

❹ Use the left vanishing point to draw a line from 7 to 3 and 8 to 4.

❺ Use the right vanishing point to draw a line from 7 to 5 and 8 to 6.

❻ Continue on the next page and repeat these steps.

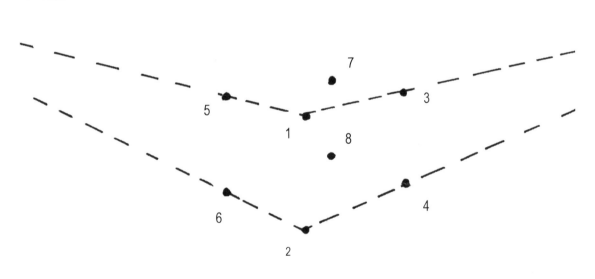

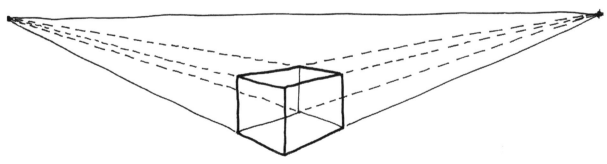

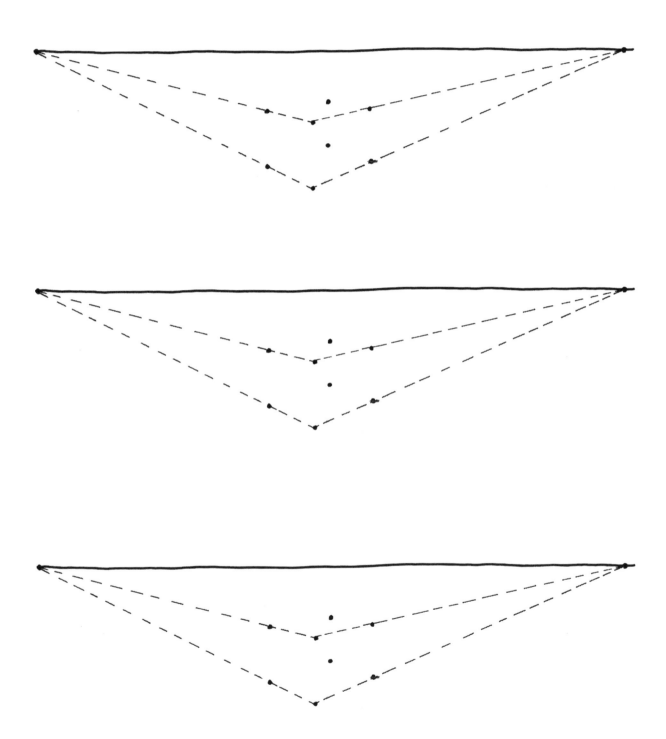

ACTIVITY 2.6   BOXES FROM DIFFERENT VIEW POINTS

In this activity, you will be drawing two-point perspective boxes from different viewpoints. Each box starts with a leading edge, perspective lines come from the vanishing points, and vertical lines are parallel to the leading edge. Use the directions on this page and complete your activity on the next page.

❶ Before starting your own drawing, add guidelines to the drawing below using your straight edge. Add the guidelines from the vanishing point to each box. Notice each box has a leading edge and the angle lines for the side of the box comes from a vanishing point. The top of the box has perspective lines that cross in the back.

❷ On the next page, turn your book horizontally, draw the horizon line at the top and add vanishing points on each end.

❸ Draw your own boxes by starting with a leading edge. Use your straight edge tool to make the guidelines from the vanishing points.

❹ Use perspective guidelines to draw the top and bottom lines.

❺ Use parallel lines to finish the sides of the box.

❻ Draw boxes below, above and on the horizon line.

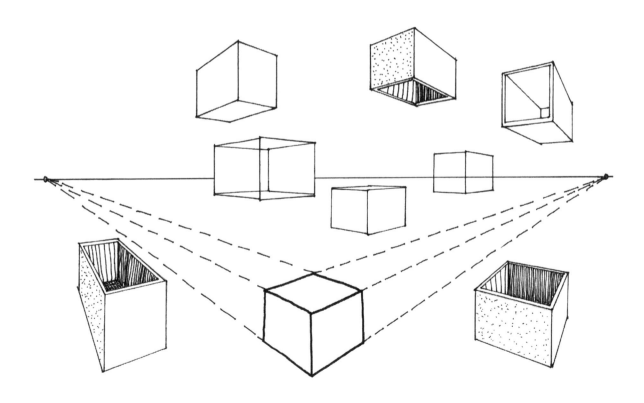

ACTIVITY 2.7   FINDING PROPORTION WITH A GRID

One of the key steps to drawing is to break down complicated objects into smaller parts. This allows you to focus on a portion of the object detail and to notice the relationship of shapes to each other.

In this activity, use the picture of the match box top as a guide to practice the divide-the-square technique for finding proportion. Draw the image of the match box top using the steps and examples as a guide:

❶ Measure the image and use this as a guide for your drawing. Draw your box shape twice as big as the actual image and using the same proportions.

❷ Using light pencil lines, divide your rectangle into equal divisions. Start by drawing a cross from corner-to-corner to find the center. Then divide the box into quarters by drawing a cross through the middle. Continue to divide each new square in this same manner.

❸ Notice how the divided rectangles provide a proportion guide for the shapes on the match box top. Lightly add the basic shapes of the image to your grid.

❹ Using your marker, added your final lines and details. Erase any visible pencil markings.

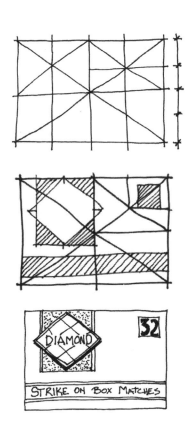

ACTIVITY 2.8 DRAWING A SOFA WITH A GRID

In this activity, using the box and outline below, draw the image of the sofa, pillows and pictures as seen in the drawing. On the outside of the box are lines providing 1/4th division of the box to use as a guide for placement of the objects. Plan to draw this free hand without the straight edge tool.

❶ Start by dividing the image to the right into equal parts by lightly drawing a pencil grid over the image.

❷ Use the rectangular shape drawn below for your drawing. Start by creating a pencil grid using the division marks on the side as a guide.

❸ Add details to the sofa outline using your marker.

❹ Use the rectangular boxes below to render each finish before adding it into the drawing. Continuing with your marker, add rendering of the two pictures.

❺ Add the pillows and the rug to your drawing using the grid as a guide for placement.

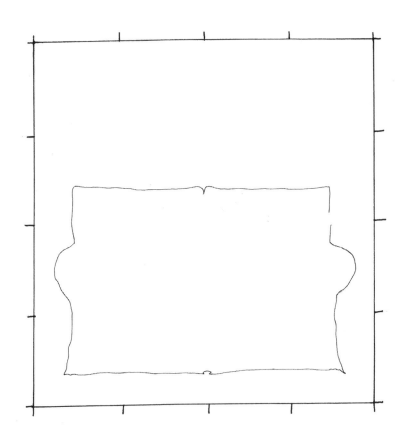

ACTIVITY 2.9   DIVIDING A RECTANGULAR SHAPE

For this activity, you will be dividing the rectangular shape into equal parts. The first rectangular shape has equally divided horizontal lines. Refer to chapter 2 of the reference book for an illustrated example of this activity.

❶ In the first box, draw a diagonal line from the bottom left corner to the top right corner.

❷ Draw a vertical line where the diagonal line intersects with the horizontal.

❸ In the second and third box, repeat the steps including dividing the shape into equal horizontal parts.

❹ Draw another set of rectangular shapes the same size and repeat the steps to dividing each of them.

ACTIVITY 2.10

ADDING RECTANGULAR SHAPES

For this activity, you will be adding to the rectangular shape in equal parts. Use the rectangular shape below to get started. Refer to chapter 2 of the reference book for an illustrated example of this activity.

❶ Using a dashed line, find the center using the "X" method.

❷ Next, draw a line from the bottom corner through the midpoint to meet the extended top line. Where these two lines intersect, draw a vertical line. These two rectangular shapes will be the same length.

❸ Repeat step #2 to continue to add more rectangular shapes.

❹ Do this activity a second time in the shape below.

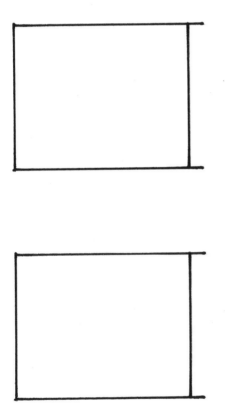

For this activity, you will be adding to the rectangular plane in perspective demonstrating the concept of foreshortening. You will be using the same steps to add of additional shapes from the last activity. This time the shapes will be getting smaller as they move away from the front, this concept is foreshortening. Use the rectangular plans below to get started. Refer to chapter 2 of the reference book for an illustrated example of this activity.

❶ Start with the rectangular shape on the left. Using a dashed line with your pencil, find the center using the "X" method in the first rectangular plan.

❷ Next, draw a line from the bottom corner through the midpoint to meet the extended top line. Where these two lines intersect, draw a vertical line. These two rectangular shapes will be the same length.

❸ Repeat step #2 to continue to add more rectangular shapes.

❹ Repeat these steps with the rectangular shape on the right.

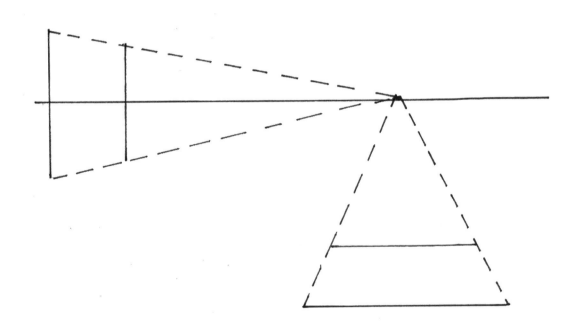

ACTIVITY 2.12 PROPORTION IN PERSPECTIVE

For this activity, use the horizon line and initial perspective lines to practice using divide-the-square technique in a perspective drawing. Plan to use a straight edge tool for this activity.

❶ Draw a square shape using the vanishing points to create your perspective lines.

❷ Draw an "X" from corner to corner inside the square shape.

❸ Draw a cross from side to side using the vanishing points and the center point of the "X" as guides.

❹ Repeat these steps to divide the square again.

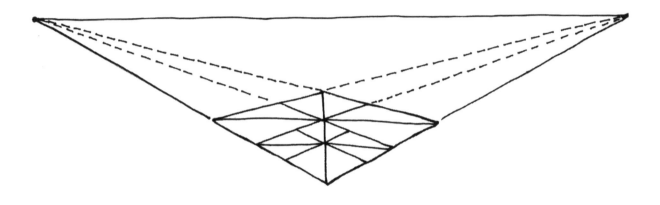

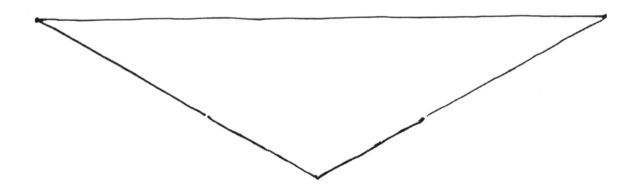

For this activity, you will be adding to the rectangular objects in perspective. You will be using the same steps to add an additional shape. This time the shape will be getting smaller which is foreshortening. Use the rectangular plans below to get started. Refer to chapter 2 of the reference book for an illustrated example of this activity.

❶ Start with the rectangular shape on the left. Using a dashed line with your pencil, find the center using the "X" method.

❷ Using the center, draw a guideline from the bottom left corner through the center. Where this intersects with the top line is the next with of the box.

❸ Draw more receding boxes using the same steps.

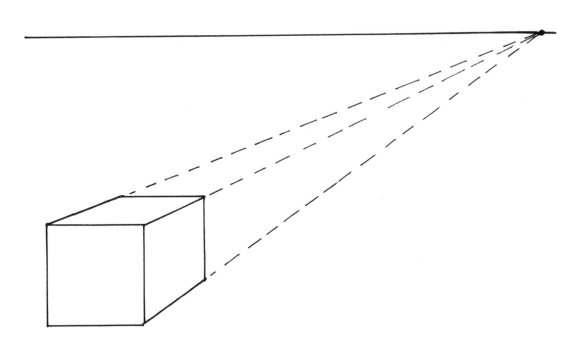

In this last chapter activity, you will be incorporating all the techniques from Chapter 2 using an everyday item of your choosing. Use the example drawn and the instructions to complete the activity on the next page. Here are a few ideas for objects that you could draw: tissue box, toaster, decorative box, toolbox, jewelry box, phone book, frame, remote control or make-up compact.

❶ Find a simple everyday item and be sure it is a *rectangular* shape.

❷ Start your drawing using your pencil, with a simple 2-point perspective box. Remember to start with a leading edge and use your straight edge to draw perspective lines using "imaginary" vanishing points. Plan to have your drawing measure at least 2 or 3 inches in size.

❸ Add guidelines and shapes outlining the details of your item.

❹ After the outlines are drawn, use marker to redraw the contour shape. Then, erase any visible pencil lines.

❺ Add the details, textures and patterns to complete your drawing. Use your pencil to create guidelines or guide marks to aid with adding the details. You can alternate between pencil and markers and erase as you go.

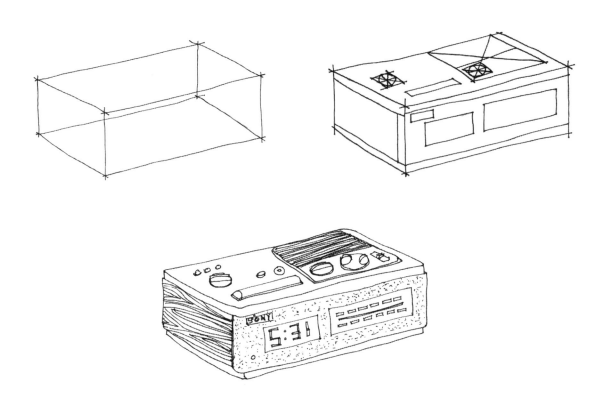

Chapter 3

Cylinders

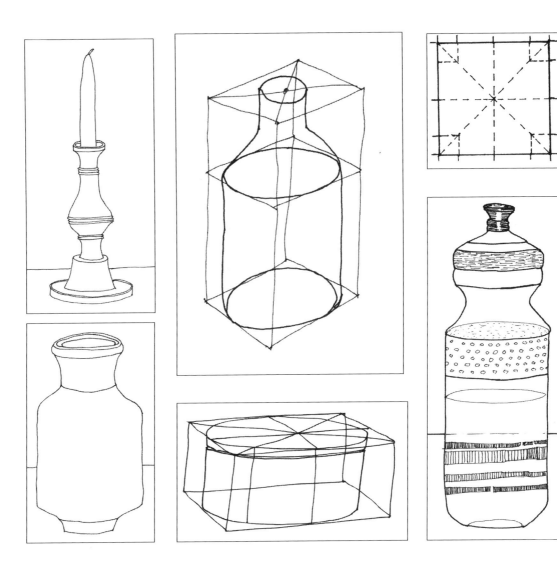

ACTIVITY 3.1 DRAWING CIRCLES & ELLIPSES

For this activity, use the box shapes below to practice the steps for drawing circles and ellipses. Refer to Chapter 3 of the reference book for further information. For each row, use the following instructions:

❶ Draw a diagonal line from each corner to create an "X".

❷ Draw a cross thru the center of the box to the edges.

❸ Create guide points by splitting each "X" into thirds and marking the outside third with a guide point.

❹ Mark a dot at the end of each cross where it touches the box edge.

❺ Using the guide points created, draw curved lines from point to point to create the circle or ellipse.

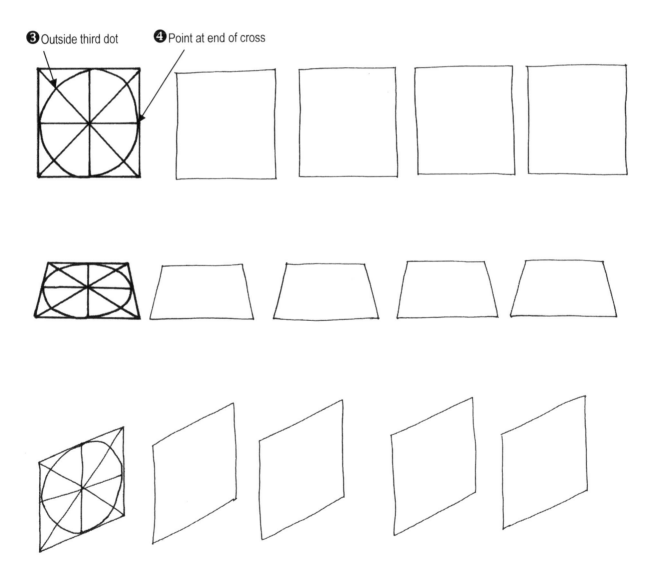

❸ Outside third dot   ❹ Point at end of cross

ACTIVITY 3.2   CYLINDER FROM DIVIDE-THE-BOX TECHNIQUE

In this activity, we will expand upon the divide-the-box technique to create cylinders. Use the box shapes below, use the instructions to draw cylinder shapes that are upright and on their sides.

❶ Divide the top and bottom of the box with an "X" and a cross using the technique from Activity 3.1. Mark your guide points.

❷ Using the guide points, draw the top of the cylinder by drawing two circular lines.

❸ Draw the bottom of the cylinder by drawing only the portion of the circle that is visible from the front.

❹ Draw the sides of the cylinder by adding straight lines connecting the top and the bottom.

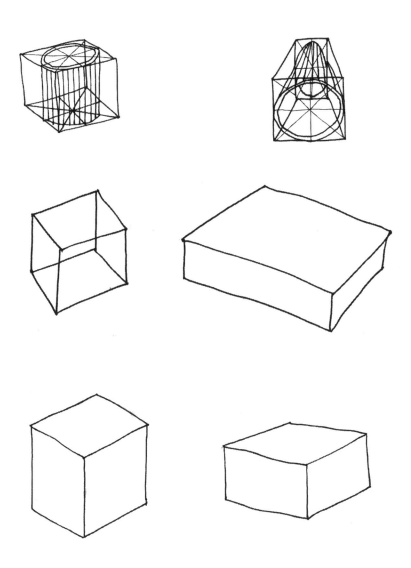

ACTIVITY 3.3   CYLINDER OBJECTS FROM THE BOX

This activity gives you an opportunity to practice drawing cylinder shaped objects. Use the drawing on the right as a guide.

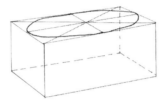

❶ Start with one of the empty rectangular box shapes below and use guidelines to divide the top.

❷ Use the guidelines to add the elliptical shape on top.

❸ Draw the bottom, front edge of the cylinder by drawing a curved line that is parallel to the top edge.

❹ Extend vertical lines from the top to the bottom to form the sides.

❺ Use your drawing marker to outline the contour or outside shape of the container.

❻ Repeat this activity using the other empty box provided and then using your own boxes.

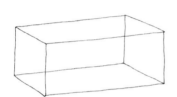
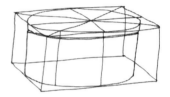

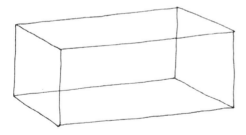
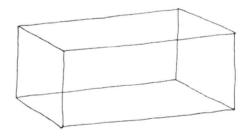

ACTIVITY 3.4 MULTI-ELLIPSE CYLINDERS

A water bottle is an example of a cylinder shaped object that contains multiple size cylinders. Use the three drawings provided as a guide and the rectangular shaped boxes below to draw a water bottle.

❶ Using your pencil, divide the first rectangular shape box on the top, in the middle and on the bottom.

❷ Draw the elliptical shapes for the top, middle and bottom.

❸ Use the edges of elliptical shapes as guides to add lines that will define the outside edge.

❹ With your marker, outline the bottle and add the label, water line and top details.

❺ Repeat this activity again using the second rectangular shaped box.

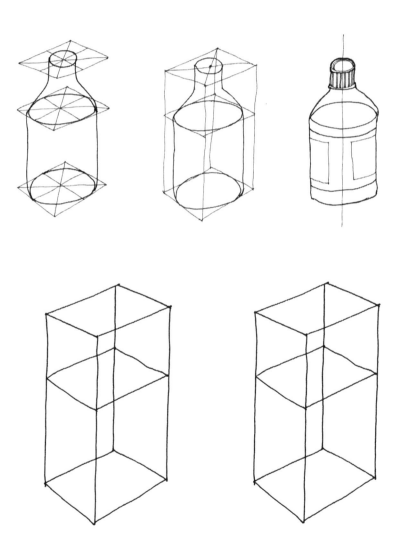

In this activity, use the space below to draw a mug. Use the two drawings provided and the example in Chapter 3 of the reference book as a guide. The first drawing demonstrates using guidelines to drawing the cylinder. The second drawing includes the handle of the mug. Here are the steps to follow:

❶ Start with a two-point perspective box and then divide the top with an "X" and a cross.

❷ Use these guidelines to complete your ellipse. Remember to draw two lines to form the top rim of the mug.

❸ Add a curved bottom parallel to the curved top.

❹ Start your handle with a side plane that is in the shape of a backward "C". This plane is completely visible.

❺ Add the outside plane that is visible from the top and side, but moves under the handle toward the bottom.

❻ Add the inside plane, which can only be seen toward the top and toward the bottom.

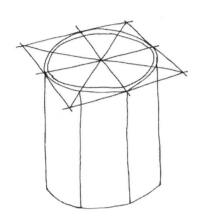

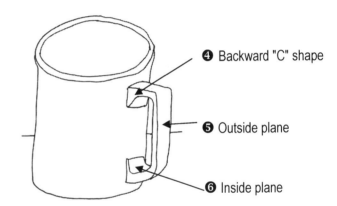

❹ Backward "C" shape

❺ Outside plane

❻ Inside plane

ACTIVITY 3.6   CYLINDERS USING CENTER LINE TECHNIQUE

For this activity, use the two drawings below to practice the centerline technique when drawing a cylindrical object. Here are the steps to use:

❶ Finish the candle drawings below using a centerline and horizontal guidelines.

❷ Draw your own candlestick to the right of the drawing. Start by drawing a centerline and then add the horizontal guidelines to assist with your drawing.

❸ Repeat these steps with the jar drawing.

This activity provides an opportunity to draw a preliminary planning sketch of cylinder shaped objects.

❶ Plan to use the space below randomly drawing cylinder shapes and objects.

❷ You can draw these shapes and objects using a rectangular box and divide the top and bottom for guidelines; or draw a center line and add guide lines to show the shape of the object. Use the drawing technique that is most comfortable for you.

❸ As you are drawing, plan to think about an idea for a theme to use in the next drawing activity.

ACTIVITY 3.8

This activity provides an opportunity to design and draw a finished image using cylinder shapes.

❶ From the practice page, choose several cylinder shapes together to create a finished image.

❷ Use your pencil to draw your shapes and objects on the page.

❸ Plan to review your composition and make changes as needed.

❹ Finish your drawing with your markers and erase your guidelines.

.

Chapter 4

Texture & Pattern

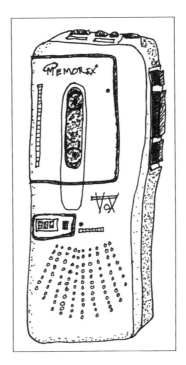

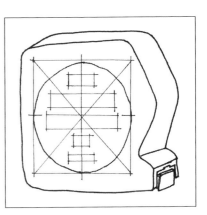

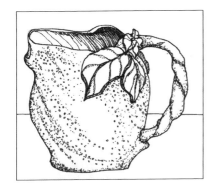

ACTIVITY 4.1   VALUE SCALES

To start this chapter of activities practice by drawing value scales using three different rendering techniques. Use your drawing markers to complete this activity. Refer to chapter 4 of the reference book for additional information.

❶ Add three additional 1" squares per row, with 1/2" space between each square, to complete the three rows below. Use your small 12" plastic T-square as a tool to draw pencil guide lines for the squares and make them each the same size.

❷ Add gradated values to the inside of each square creating a value scale. Start with parallel lines in the first row, stippling in the second row and cross-hatching in the third row.

❸ Border the square with your thickest marker to complete the activity.

ACTIVITY 4.2   ACTUAL & IMPLIED TEXTURES

For this activity, draw a variety of actual and implied textures with your markers. Use this activity to experiment with different ways to render texture. These implied textures are good references for other activities as you add texture to your drawings.

❶ Add three additional 1" squares per row, with 1/2" space between each square, to complete the three rows below. Use your small 12" plastic T-square as a tool to draw pencil guide lines for the squares, making them each the same size.

❷ Fill each new square with a rendered implied texture using your drawing marker.

❸ Border the square with your thickest marker.

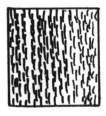

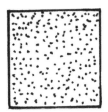

ACTIVITY 4.3 ADDING VALUE CONTRAST

In this activity, you will be practicing drawing contrasting value with a variety of patterns and textures. These will be design drawing patterns and textures. Refer to chapter 4 of the reference book for an illustrated example of this activity.

❶ Plan to have one drawing with a darker background with lighter objects on the table with the other drawing the opposite.

❷ Using your drawing marker, add a variety of pattern and textures to each element in the drawing using design drawing.

❸ Remember you will most successful with drawing a wide range of values with darks and lights.

ACTIVITY 4.4 ADDING PATTERNS TO PILLOWS

In this activity, you will be practicing adding pattern and texture to the pillow line drawing below. Refer to chapter 4 of the reference book for an illustrated example of this activity.

❶ Research and have images of the fabric that you will render as a visual reference.

❷ With your drawing markers, add different patterns and textures to each pillow shape.

❸ Remember it is important to render your dark and light values to show your texture and patterns

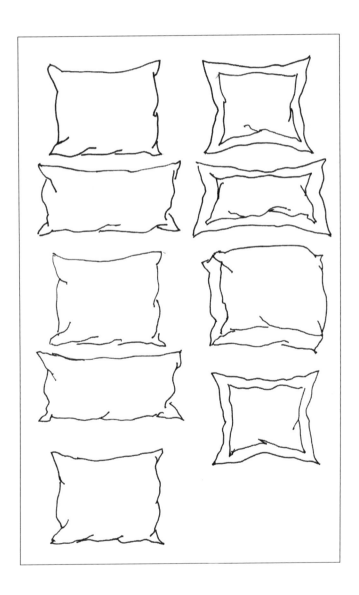

ACTIVITY 4.5   RENDERING FLOOR PLAN FINISHES

In this activity, you will be rendering finishes with a variety of patterns and textures to the floor plan line drawing below. You will be adding both flooring and furniture finishes. Refer to chapter 4 of the reference book for an illustrated example of this activity.

❶ Plan to have actual samples or images of the materials that you are rendering for visual reference.

❷ Using your drawing markers, render the different floor finishes and materials on the furniture.

❸ Draw contrasting values to the finishes that are close to each other.

❹ Darken the walls with parallel lines.

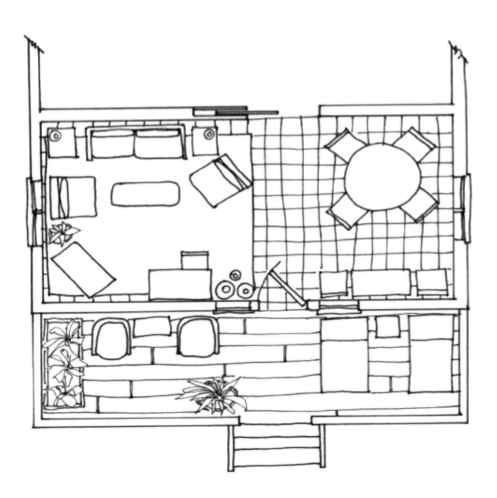

ACTIVITY 4.6 WOOD RENDERING

Before you start this activity, find two or three different wood samples to view while drawing your wood renderings. Make an effort to have the characteristics of each wood pattern be unique. The sample drawings below are plastic laminate samples. The types of wood represented are river cherry, empire mahogany and limber maple. They demonstrate a variety of unique wood patterns and values.

❶ Draw, in marker, a rectangular box shape to render each of your wood samples. If you are using plastic laminate samples, you can put your sample right on the page and trace this shape.

❷ Look closely at each sample to discover the unique characteristics of each type of wood. Over-emphasize these characteristics and values. Make the darks darker and the lights lighter than what you actually see.

❸ Draw one of the examples shown and then add your own two or three wood samples to complete the assignment.

  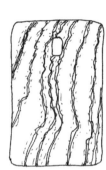

For this activity, use a rectangular shaped object made of wood that you can draw for this activity.

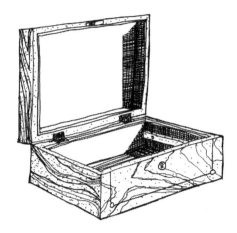

❶ Use your pencil to start your drawing.

❷ Draw the contour shape of the object in perspective and proportion.

❸ Use lines and dots to simulate the pattern and texture of wood.

❹ Over-emphasize the value and wood texture. Add as much interest as you can with the rendering of the wood.

❺ Add marker when you have completed the shape of your object and then erase the pencil guidelines.

In this activity, you will be rendering finishes with a variety of patterns and textures in the line drawing below. Decide where your light and dark values will go with the finishes that overlap each other. You will also want to have contrasting patterns with the finishes. For example, will the desk be a dark value and the walls a lighter value? How will the chair have a contrasting texture and pattern from the desk and floor finishes? Use the rectangular boxes below to render each finish before adding it into the drawing. Refer to chapter 4 of the reference book for an illustrated example of this activity.

Chapter 5

Shade & Shadow

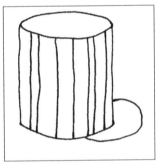

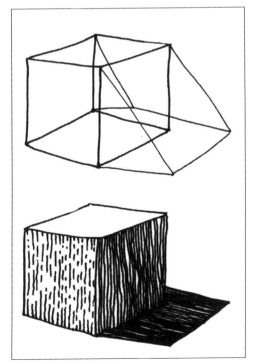

ACTIVITY 5.1   SHADE AND SHADOW WITH AN OPEN BOX

For this activity, use the large box drawing below to add shade and shadow as shown in the example on the right side. Here are the steps to follow:

❶ Keep the right side of the box white.

❷ For the inside of the box, using vertical lines, add a darker value on the right and lighter value to the left.

❸ On the outer left side of the box, using pencil, add two light horizontal guidelines, parallel to the top, which will be guidelines for value change. The first line will be half way and the second line will be one quarter the way up from the bottom.

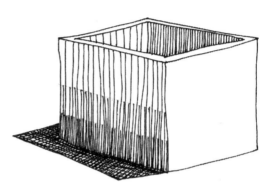

❹ For the gradation on the left side, draw lines from top and move down to bottom that have space in between.

❺ Add lines in between that are half way up the box.

❻ Add lines in between again that will be a quarter of the way up the box.

❼ Use cross-hatching for the shadow and make this the darkest value. Parallel the lines in the shadow with the outside contour lines.

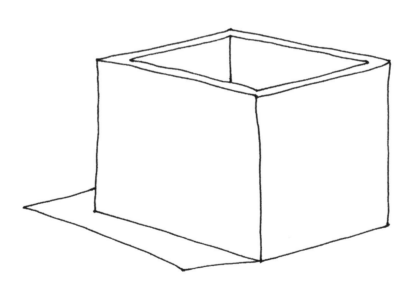

In this activity, you will be adding shade and shadow to simple geometric shapes. For each shape, add both the shading and the shadow. Use the example in Chapter 5 of the reference book as a guide.

❶ For the first shape on the left, use your marker to add gradated value using <u>stippling</u>.

❷ For the second shape, again use your marker to add gradated value, except this time use <u>lines</u> to create shade and shadow.

❸ For the third shape, again use your marker to add gradated value, using your own method for create shade and shadow.

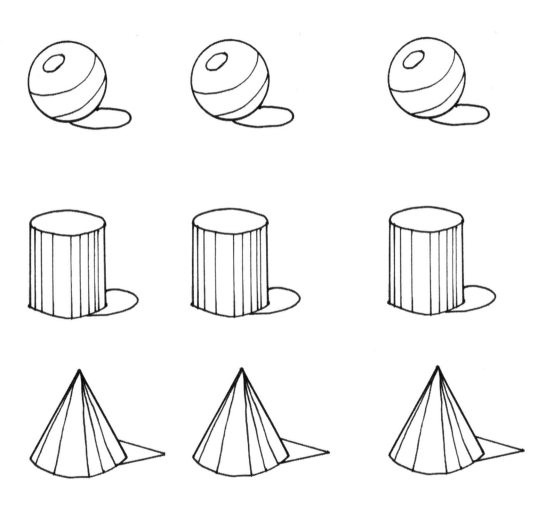

ACTIVITY 5.3   LIGHT SOURCE PARALLEL TO HORIZON LINE

In this activity, you will be adding shade and shadow to the transparent two-point perspective box below. Refer to chapter 5 of the reference book for an illustrated example of this activity. A finished example of this activity is on the right.

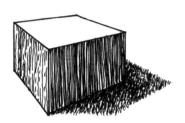

Use the box image below, which has first drawing steps with the two-point perspective box are completed. In this drawing, there are ground shadow lines from the box that start from the front corner parallel to the horizon line. The next line is from the bottom centerline and the third line is from the back corner. Each of these have estimated lengths.

Here are the steps for you to take to add the ground shadow and value:

❶ Start with the angle of the light that is a dashed line from each of the three top corners.

❷ Use a 45 degree triangle to draw these lines.

❸ Draw lines from the vanishing points that connect the points.

❹ Use vertical lines to add value to the sides of the box.

❺ Add a dark value to the shadow area using a criss-cross pattern.

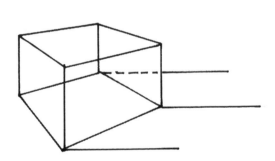

ACTIVITY 5.4 — LIGHT SOURCE IN FRONT OF VIEWER

In this activity, you will be adding shade and shadow to geometric shapes using a light source that is in front of the viewer. Refer to chapter 5 of the reference book for an illustrated example of this activity. A finished image is on the right side.

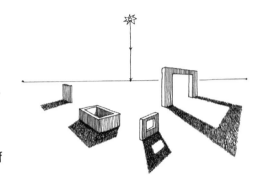

❶ Draw dashed guidelines from the light source to the top corner of each object to the ground with a straight edge.

❷ Draw solid guidelines from the point directly below the light source to the ground points of each object extending out in front of the object.

❸ Lines from the left and right vanishing points determine the back line on the ground shadow.

❹ After determining the shape of the shadow, add a dark value using a criss-cross pattern in the shadow shape.

❺ Add vertical lines to the back of the objects and inside the box shape.

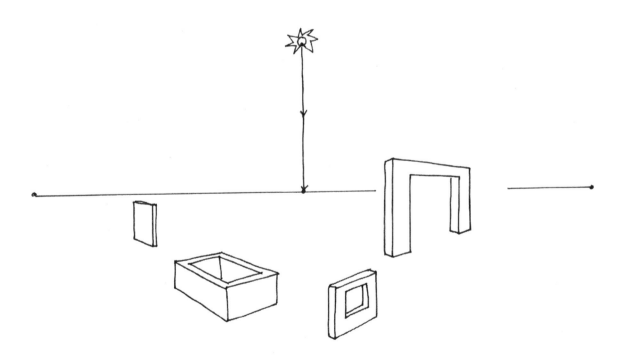

ACTIVITY 5.5

Use the photograph and rendered drawing of the brick bench below as a guide for drawing the same image in the space below.

❶ Start with your pencil to draw the brick bench shape.

❷ Add the shadow shape.

❸ Finalize the drawing in marker.

❹ Take time to render the bricks. Draw the dots to add texture. Draw these at the edge of each brick leaving the center with a lighter value. Notice how the left side has more value in the rendering since this side of the brick bench is darker.

ACTIVITY 5.6

In this activity, you will be adding shade and shadow to geometric shapes using a central light source. Use the interior image below to complete the assignment. Refer to chapter 5 of the reference book for an illustrated example of this activity. A finished image is on the right.

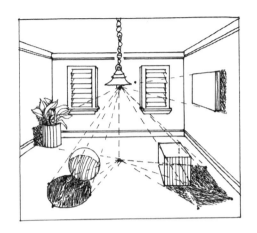

❶ Draw dashed guidelines from the light source to the top corner of each object and extended to the ground.

❷ Draw solid guidelines from the point directly below the light source, views vanishing point, to the ground points of each object.

❸ Lines from the left and right vanishing points determine the back line on the ground shadow.

❹ After determining the shape of the shadow, add a dark value using a crisscross pattern in the shadow shape.

❺ Add vertical lines to the back of the objects.

ACTIVITY 5.7

In this activity, you will be adding shade and shadow to the items on the table. Use the lamp as the light source where the "X" is located on the shade. Refer to chapter 5 of the reference book for an illustrated example of this activity.

❶ In the image below, start with pencil guidelines and finish your drawing with drawing markers.

❷ Use a straight edge and pencil from the "X" to the objects to make your shadow guidelines.

❸ Add rendering to the image using a variety of textures and pattern. The areas on the items that are shaded and in shadow will need a darker pattern.

Chapter 6

Furniture

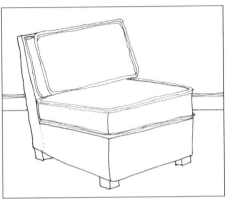

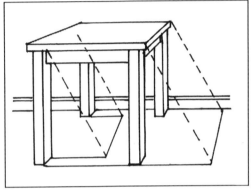

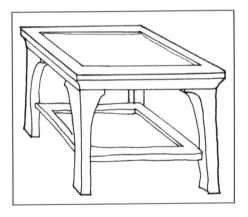

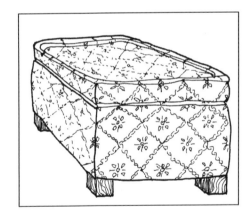

For this activity, use the illustration of the finished one-point perspective side table as a guide for completing your drawing. With your pencil, extend the sides of the example finding the center point where the angles intersect. Use this as a guide for the vanishing point for the table drawing that you start using the elevation. Review the example in Chapter 6 as a reference of this activity.

❶ Using your pencil, add the molding edge below the tabletop.

❷ Next, add the two stiles on the inside box, the top rail, and drawer rail.

❸ Refine the base shape using guidelines.

❹ Find the center of the drawer by drawing an "X" from corner to corner; then add the knob shape.

❺ Add a lightly drawn horizon line above the elevation drawing. Place this at a distance above the elevation that is the same height as the front of the elevation.

❻ Locate the center of the elevation and use this for the vanishing point on the horizon line. Draw the top perspective guidelines from the vanishing point to each corner of the elevation.

❼ Use half of the height of the elevation for the depth of the tabletop.

❽ Using your wide drawing marker pen, redraw the dresser outline right on top of the pencil. For the other lines, use a drawing marker with a smaller width. Erase remaining light pencil lines.

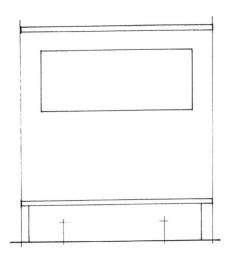

ACTIVITY 6.2   CASE GOODS IN ONE-POINT PERSPECTIVE

For this activity, use the elevation to finish drawing one-point perspective tables and add rendered finishes. Refer to chapter 6 of the reference book for an illustrated example of this activity.

❶ The table on the left shows the right side of the legs and the table on the right shows the left side of the legs.

❷ The middle table shows a flat front drawer with the shelf above receding back with the vanishing point as a guide.

❸ The wood and stone patterns on the top of the tables use the vanishing point for vertical lines found in the pattern.

ACTIVITY 6.3   SIDE TABLE ADDING SHADE AND SHADOW

For this activity, use the illustration of the one-point perspective side table is a guide. Plan to add shade and shadow to the table below. Refer to chapter 6 of the reference book for an illustrated example of this activity.

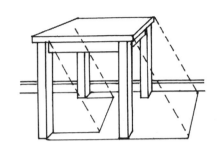

❶ Draw the ground shadow lines from the front corner leg parallel to the horizon line.

❷ Repeat this with the middle corner leg and back corner leg. Draw the bottom of the leg to provide this corner.

❸ Start with the angle of the light that is a dashed line from each of the three top corners.

❹ Use a 45 degree triangle to draw these lines.

❺ Draw lines connecting the intersections of the ground lines and light angles.

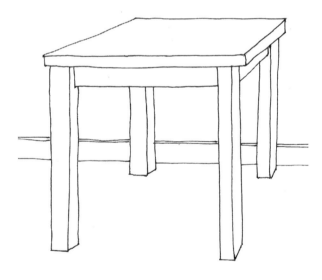

For this activity, practice adding surface rendering to each of the side tables below. Plan to add wood to the first table and your own choice of a finish to the second table. Refer to chapter 6 of the reference book for an illustrated example of this activity.

❶ Research and find images of the two finish materials that you will render as visual references.

❷ Start your rendering pattern with light pencil using line and stippling to imitate the surface material. Remember to use the angles of the tabletops as a guide for adding your patterns. Draw your patterns in perspective.

❸ With your drawing markers, add the final rendering patterns.

❹ Plan to have the light source from the left side. Add a darker value pattern to the right side of the tables.

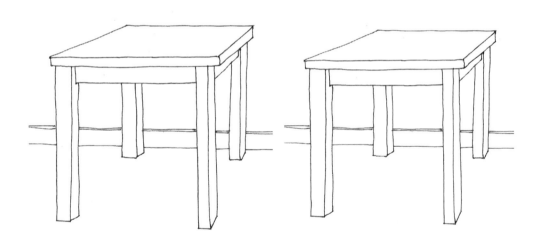

ACTIVITY 6.5 ONE-POINT PERSPECTIVE SIDE CHAIR

In this activity, practice drawing a one-point perspective side chair using an imaginary vanishing point that is on the right side of the chair. On the right, there is an example of a finished chair. The chair below is an example of the first steps taken to draw a one-point perspective chair. Plan to find an actual simple rectilinear side chair to draw. This will give an opportunity to practice drawing the unique features of a chair.

❶ Start with very light pencil and draw a one-point box shape for the base next to the example provided to assist with the size and proportions. Draw the back right leg and seat box shape.

❷ Add another thin box shape to form the back of the chair.

❸ Using a darker pencil, refine the shape of the chair and add more details including the curve in the back of the chair, the legs and base railing.

❹ To finish the drawing, use your markers to add contour lines and rendering.

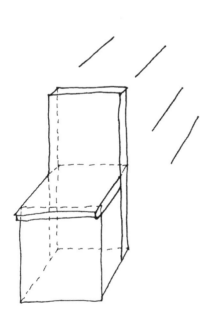

For this activity, use the instructions on this page and draw the activity on the next page.

❶ Start by drawing a horizon line on the top of the page and placing a vanishing point in the center.

❷ Add the center ottoman first. Draw a flat front rectangle using your pencil, straight edge and T-square tools.

❸ Use the vanishing point to establish the topsides and then draw a horizontal line for the top back.

❹ Add rectangular flat front shapes for the feet, again using the vanishing point to find the properly angled lines for the feet.

❺ Use the vanishing point and horizontal lines to establish the location of the four corners for the buttons.

❻ Add the next two ottomans, again starting with a flat front rectangular shape to create the front.

❼ Use the same single vanishing point to establish the top, the side panel and the sides of the legs. Notice how the visible side panel changes with the position of the ottoman relative to the vanishing point.

❽ Add your drawing marker on top of the pencil to add detail and to complete the image.

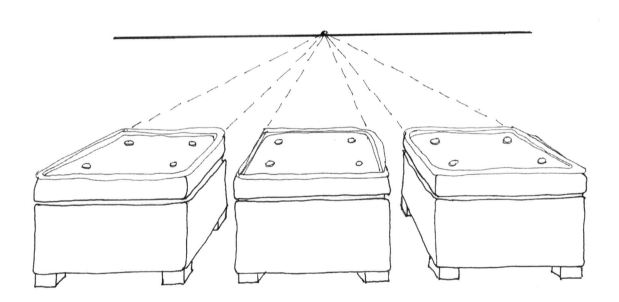

ONE-POINT PERSPECTIVE OTTOMAN

ACTIVITY 6.7   RENDERING FABRIC PATTERN

Before starting this activity, research and have an image of the fabric that you will render as visual reference. You will need to see enough of the fabric pattern design to draw it into the grid. Review the example in Chapter 6 as a reference of this activity.

❶ Start by using the square shape with the grid to draw the pattern repeated flat. Image your pattern as a yard of fabric that you are looking at rolled out on a table. Draw the pattern motif in each square. You can lightly add pencil grid lines inside the square if this is helpful.

❷ Use the pattern drawn out on the grid as your guide for adding pattern to the ottoman.

❸ Add the pattern into the grid on the ottoman. Keep in mind that the further away the pattern is, the less detail you will need to add.

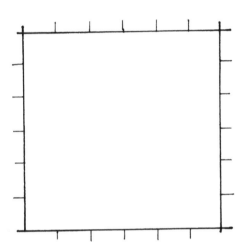

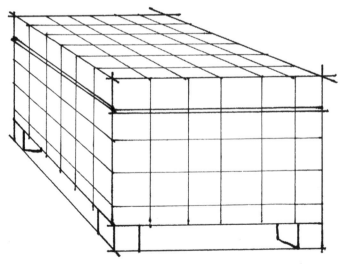

For this activity, add rendered finishes to the line drawing below. Research and have images of the different finishes that you will render as visual reference. Start by drawing your rendered finishes in the shapes below. Remember to organize your dark and light values on the finishes that overlap each other. Review the example in Chapter 6 as a reference of this activity.

ACTIVITY 6.9   TWO-POINT PERSPECTIVE SIDE TABLE

For this activity, use the finished two-point perspective side table drawing on the right as a guide for completing the drawing below. Use in chapter 6 the furniture anatomy information as a reference for this activity.

❶ Start with your pencil to divide the front plane of the piece with an "X" to locate the middle.

❷ Draw the middle panel and top panel.

❸ Divide the contour lines of the drawer and locate the center to draw the knob shape.

❹ Draw the molding on the left side panel. Remember to add the inside depth on the side and bottom.

❺ Add the books on the lower shelf.

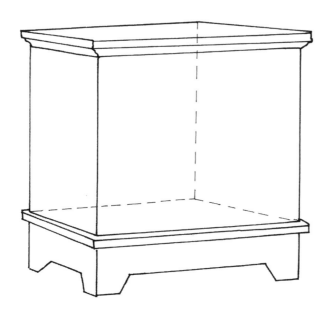

ACTIVITY 6.10   TWO-POINT PERSPECTIVE TABLE LEGS

In this activity, practice drawing two table legs using the correct vanishing point using the drawing below. The drawing on the right demonstrates how the legs on the dark value side of the table uses the right vanishing point and the lighter value side of the table uses the left vanishing points. Review the information in Chapter 6 as a reference of this activity.

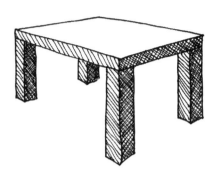

❶ Start with drawing the right tabletop with the right vanishing point and the left table topline with the left vanishing point.

❷ Draw each of the legs using the correct vanishing points for each of the different sides.

❸ Add value using a diagonal line show the darker and lighter sides.

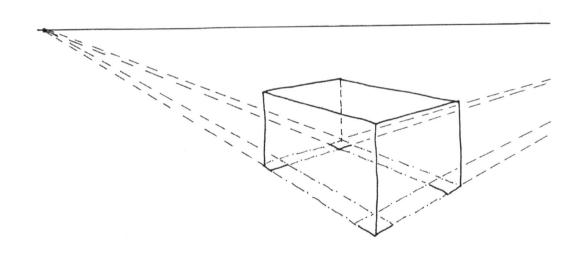

ACTIVITY 6.11   TWO-POINT PERSPECTIVE BOOKCASE

For this activity, use the instructions on this page and draw the activity on the next page. Provide on the next page is the shell of the bookcase with the horizon line and vanishing points. Refer to chapter 6 of the reference book for an illustrated example of this activity.

Top molding:

❶ Draw a perspective guideline from the (LVP) & (RVP) to add the top molding.

❷ Where the two guidelines intersect, add a curved molding line from the top left corner line to the molding line.

❸ Repeat this curved molding line on the right side from the bookcase top corner to the molding line. Use your eye to determine the length of the molding

❹ Add a third molding line in the back right corner.

Bottom base:

❶ Draw two parallel lines from the RVP and two parallel lines from the LVP. Where they intersect, draw a horizontal line representing the bottom right front corner.

❷ Draw the inside top with a perspective line from the RVP to the left front corner of the bookcase.

❸ Inside top, draw a second perspective line from the LVP to the right corner.

Shelving Depth:

❶ Draw a vertical line from the inside top left corner down to the bottom inside left corner.

❷ Use the perspective guidelines from the shelf front to the RVP to determine the shelves.

TWO-POINT PERSPECTIVE BOOKCASE

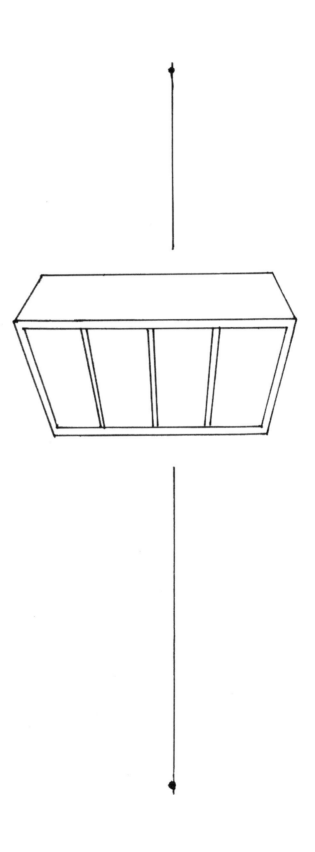

ACTIVITY 6.12

TWO-POINT PERSPECTIVE SIDE CHAIR

In this activity, practice drawing a two-point perspective side chair using imaginary vanishing points.   On the right, there is an example of a finished chair. Below is an example of the first drawing steps. For the drawing, plan to find an actual simple rectilinear side chair or use the same chair from activity 6.4.

❶   Start with very light pencil and draw a two-point box shape for the base. Draw this box shape next to the example to assist with your size and proportion. Draw the back front leg and seat box shape.

❷ Add another thin box shape to form the back of the chair.

❸ Using a darker pencil, refine the shape of the chair and add more details including the curve in the back of the chair and the legs and base railing.

❹ To finish the drawing, use your markers to add contour lines and rendering.

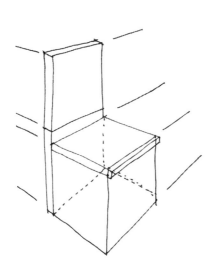

For the first this activity, use the instructions on this page and draw the activity on the next page. Plan to turn the book horizontally to provide extra space for the two vanishing points.

❶ On the next page, draw a horizontal line, across the whole page, and add two vanishing points on either end of the line.

❷ Start the ottoman drawing by first drawing the vertical leading edge line using your pencil. Plan to have this close to the bottom of the page.

❸ Use the vanishing points to establish the shape of the base, both the sides, and the top.

❹ Using the vanishing points, add double lines to define the cushion top.

❺ Add the feet below the base.

❻ Add piping to the top and guidelines from the vanishing point draw the buttons.

❼ Add your drawing marker on top of the pencil to complete the image.

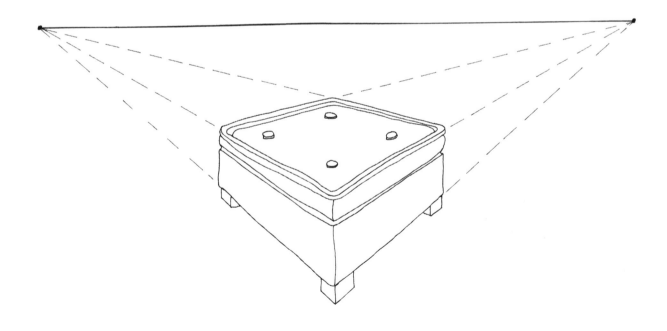

Before starting this activity, research and have two images of the fabric that you will render as visual reference. You will need to see enough of the fabric pattern design to draw it into the grid. Review the example in Chapter 6 as a reference of this activity.

❶ Start by using the square shape with the grid to draw the pattern repeated flat. Image your pattern as a yard of fabric that you are looking at. Draw the pattern motif in each square. You can lightly add pencil grid lines inside the square if this is helpful. This will show you how the pattern is organized.

❷ Use the pattern drawn out on the grid as your guide for adding pattern to the first chair.

❸ In the second chair, with your pencil and the grid shape to the chair and repeat the steps above.

For this activity, add rendered finishes to the line drawing below. This interior space is a Connie Riik design. Research and have images of the different finishes that you will render as visual reference. Start by drawing your rendered finishes in the shapes below. Remember to organize your dark and light values on the finishes that overlap each other. Review the example in Chapter 6 as a reference of this activity.

Chapter 7

Interior Details

ACTIVITY 7.1 ADD ACCESSORIES TO BOOKCASE

For this activity, use the finished two-point perspective bookcase below and draw a variety of accessories on each shelf. Plan to balance each shelf with the decorative items. Have the visually heavier items on the bottom of the bookshelf. Use picture frames and plants. Add texture and pattern. Review the example in Chapter 7 as a reference of this activity.

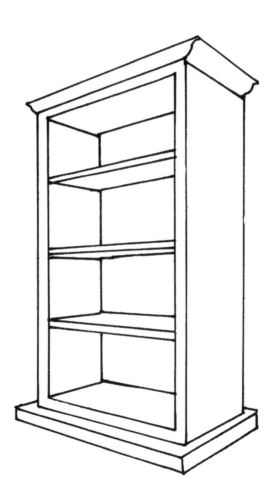

This activity will give you an opportunity to practice rendering art. Before starting, have four different image examples that you can use as a starting point for each drawing. The goal is to render each art piece in a subtle, suggested impression of the actual picture. Use the four frames below to draw your images. Here are techniques that you can use:

❶ Create an impression of the image with low values using your fine line drawing marker.

❷ Avoid using symbols which attract the eye and will create an over emphasized focal point. These shapes include circles, triangles, crosses or spirals.

❸ Avoid drawing large single objects such as a flowers or trees.

❹ Add a matte to several of your drawings. Add the depth line for the matte on top and to the right side.

The goal of this activity is to practice drawing lamps with the correct proportion. For these drawings, use the center line technique covered in Chapter 3. Plan to redraw each lamp on the right side. Use the dimension lines as a guide for drawing the correct proportions.

❶ Start with pencil and draw the dimension line horizontally to the right side matching the example.

❷ Draw a center guideline for the middle of the lamp and use this plus the dimension line to define proportion.

❸ Draw vertical guidelines for the width of the lamp. Connect these lines to form the contour of the lamp.

❹ Add the remaining details then complete the drawing by adding your markers.

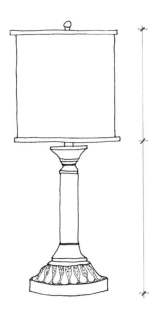

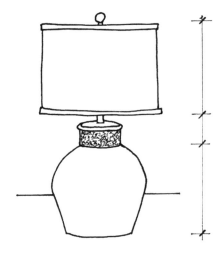

For this activity, find a tabletop picture frame that you can use to draw in a two-point perspective. Position the frame with the left side as a leading edge. Use the drawing below as a guide.

❶ Start by drawing the rectangular front of the frame.

❷ Add an additional rectangle on the inside to form the molding lines. Create rendering of depth to the right side and bottom.

❸ Add the outside frame depth to the left side and the top. The visible depth will depend on the position of your frame.

❹ Notice that this example has two layers of molding. Look closely at your frame and any unique characteristics.

❺ Add a darker value to the sides of the frame to render shading.

❻ Add a rendered image.

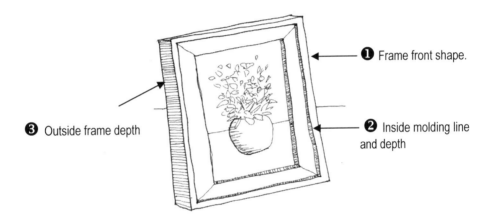

❸ Outside frame depth

❶ Frame front shape.

❷ Inside molding line and depth

For this activity, you will focus on a single leaf. Take a leaf off a plant and study its unique characteristics. You will be drawing this leaf three times, imitating the example shown. Start the drawing with pencil guidelines as needed and finish it with drawing markers.

❶ In the first drawing, use line to draw the contour of the leaf. Use two lines to add the veins on the inside of the leaf.

❷ In the second drawing, repeat the first drawing and add stippling for value and shape. Draw the stippling in clusters close to the veins giving the leaf a curved shape.

❸ In the third drawing, repeat the steps of the first and second drawings. Add a rectangular shape behind the leaf and draw value with vertical lines.

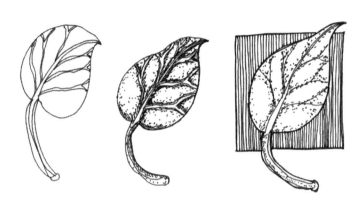

This activity provides basic steps for creating a simple container, an integral part of indoor plants. Use guidelines to develop the plant container. This is shown in the top drawing demonstrates. The bottom drawing shows the finished plant container. Plan to draw your container on the right side of the finished drawing. Here are the steps you can use.

❶ In pencil, start the container with a rectangular two-point perspective shape.

❷ Use the "X" and the cross from corner to corner to find the center. Draw a horizontal line through the center.

❸ Draw the rim of the container with two ellipses and using your guidelines.

❹ Add the top portion of the container by first drawing a rectangle beneath your rim. Notice that the vertical lines originate at the outer edge of the rim.

❺ Add another rectangle beneath the first to form the base of the container. Add vertical curved lines from the bottom of the top rectangle to the base of the second rectangle to define the sides of the container base.

❻ Add curved lines connecting the sides of both rectangles to form the curved front edges of the container.

❼ With your markers, draw the container contour lines and erase pencil guidelines.

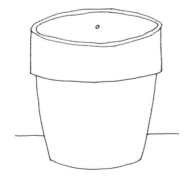

ACTIVITY 7.7   ADDING PLANT TO CONTAINER

In this activity, you will add a plant to the container. Use the container below and draw a plant using the following steps. Plan to have an image of a plant or an actual plan to use as your visual reference for this activity. Review the example in Chapter 7 as a reference of this activity.

❶ Using a pencil, light draw in the front leaves of the plant. Add several other leaves to give an overall picture of the plant.

❷ Start drawing the leaves at the front of the plant with your drawing marker. Build the back of the plant with the overlapping leaves.

❸ Draw dirt or rocks inside the container.

❹ Outline the container with your marker.

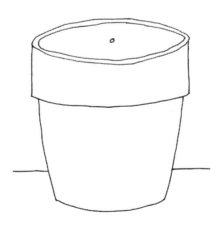

For this activity, add rendered finishes to the line drawing below. This is an interior space designed by Connie Riik. Research and have images of the different finishes that you will render as visual reference.   Start by drawing your rendered finishes in the shapes below. Remember to organize your dark and light values on the finishes that overlap each other. Review examples found in Chapter 4 as a reference of this activity.

For this activity, use the instructions on this page and draw the activity on the next page. You will note the rectangular shape is the trunk and the dot is the vanishing point. Review the example in Chapter 10 as a reference of this activity.

First Part:

❶ Draw the two sides of the sofa equal distances using the center guideline. Draw the sofa wider than the trunk and locate the base of the sofa slightly above the truck.

❷ Use the perspective guidelines to draw the depth of the arms.

❸ Draw horizontal lines indicating the sofa base and seat cushions estimating the width.

❹ Use the perspective guidelines to draw the sides of the cushions.

❺ Intersect the two side seat lines with a horizontal line.

Second Part:

❶ Finish drawing the sides and back of the arms.

❷ Divide the seat cushion in half using two lines that are from the vanishing point.

❸ Draw the back cushion rectangular shape along with drawing a sofa back on each side above the arms.

❹ Draw guidelines for the lamp and two containers shapes.

References

Gerds, Donald A, *Perspective: A Step-By-Step Guide for Mastering Perspective by using a Grid System*, DAG DESIGH, 2002

Gordon, Robert Philip, *Perspective Drawing: A Designer's Method*, Fairchild Books

Hanks, Kurt *Rapid Viz, A New Method for the Rapid Visualization of Ideas.* California: William Kaufmann, Inc., 1980

Koenig, Peter A., *Design Graphics, Drawing Techniques for Design Professionals*, Pearson Hall, 2006

Laseau, Paul, *Freehand Sketching, an Introduction.* New York W.W. Norton, 2004

Mitton, Maureen, *Interior Design Visual Presentation, A Guide to Graphics, Models, and Presentation Techniques.* New York John Wiley & Sons, Inc. 2008

Montague, John, *Basic Perspective Drawing, A Visual Guide* New York John Wiley & Sons, Inc. 2005

Natale, Christopher, *Perspective for Drawing Interior Space*, Fairchild Books, 2011

Pile, John, Perspective *for Interior Designers, Simplified Techniques for Geometric and Freehand Drawing*, Watson-Guptill Publications, 1989

Wirtz, Diana Bennett, Hand Drafting for Interior Design, Fairchild Books, 2010